AF878795

To Augusto Morello

the world in a room
When Objects Have Names of Places

© 2002 Triennale di Milano
All rights reserved

Printed in Italy
ISBN 88-8158-412-3

Edizioni della Triennale

La Triennale di Milano Foundation
viale Alemagna 6
20121 Milan
Tel. +39-0272434.1
Fax +39-0289010693
www.triennale.it

Distribution
Edizioni Charta

Via della Moscova 27
20121 Milan
Tel +39-026598098/026598200
Fax +39-026598577
Email: edcharta@tin.it
www.chartaartbooks.it

Permanent Collection of Italian Design

il mondo in una stanza
Quando gli oggetti hanno nomi di luoghi

Palazzo della Triennale
Viale Alemagna 6, Milan
October 16, 2002 – February 2, 2003

Curator
Silvana Annicchiarico

Exhibition Design
Giancarlo Basili

Graphic Design
Daniele Mastrapasqua

Organizational Support
Elisa Testori

Decoration Roberta Dondi
Installation Produzioni Reali
Animation Paolo Pelanda
Lighting Marzoratimpianti

Translation
Victoria Storke, Studio MVM, Milano

Cover
Tramonto a New York
Gaetano Pesce, 1980, Cassina

LA TRIENNALE DI MILANO

PERMANENT COLLECTION OF ITALIAN DESIGN

the world in a room

When Objects Have Names of Places

Edited by Silvana Annicchiarico

CHARTA

La Triennale di Milano Foundation

Board of Directors
Aldo Bassetti
Renato Besana
Paolo Caputo
Roberto Cecchi
Silvia Corinaldi Rusconi Clerici
Giuseppe Panza di Biumo
Davide Rampello

Board of Auditors
Alfredo Giacomazzi
Adriana Rampinelli, *President*
Domenico Salerno

General Manager
Andrea Cancellato

General Affairs
Federica Molteni
Annunciata Marinella Alberghina
Franco Romeo

Initiatives
Laura Agnesi
Roberta Sommariva

Library, Documentation, Archives
Ferruccio Dilda

Technical and Administrative Services
Alessandro Cammarata
Giuseppina Di Vito
Marina Gerosa
Isabella Micieli
Franco Olivucci
Pierantonio Ramaioli

Press Office
Antonella La Seta Catamancio

Marketing Office
Laura Benelli

Curator of the Design Permanent Collection
Silvana Annicchiarico

Scientific Contributors:
Architecture, Luca Molinari
Design, Silvana Annicchiarico
Fashion, Raimonda Riccini
International Relations, Aldo Colonetti

Producer
Roberto Bucci

Contents

Preface

After *I'm No Lady*, dedicated to Italian design objects with women's names, and *Animal House*, focusing on objects with animal names and shapes, *The World in a Room* now completes a tryptych: the three-part exhibition cycle called "Words and Things," in which La Triennale di Milano has experimented with new methodologies and interdisciplinary ways of approaching the complex and varied design universe.

Open-ended research and reflection on the rapport between objects and their names – a project met with significant public success and the flattering attention of the press – has permitted the discovery of new sights and has initiated stimulating analytic perspectives within the world of design and design production. In addition to uniquely exploiting and enlarging the Italian Design Permanent Collection, there is the hope that due to these initiatives the project to transform the Collection into a real design museum can finally take shape, as proposed and pursued during the late Professor Augusto Morello's Presidency.

La Triennale di Milano

The Register of Modernity

Pippo Ciorra
Professor, School of Architecture at Ascoli Piceno

Names of Places

"Malnome, Malpasso, Malafede, Malagrotta, Valle Oscura, Passoscuro, Fosso Sanguinara, Femminamorta, Pantano dell'Intossicata..."

Names of places, as Valerio Magrelli explained a few years ago in a noted text on Roman suburban toponymy, have a special power: "they lie in wait," they strongly condition the perception and image that we form of a place. "But where do you find names like these: a stretch of trodden blood, tilled and plowed, fallow fields of epithelial tissue, a plot of meat, lawns of skin, fat and tegument..." *Tramonto a New York* (sunset in New York), *I Sassi* (stones) of polyurethane, *Pirellone*, *Atollo* (atoll), *Louisiana*, *Faraglioni*, *Dune* (dune)... What do they have to do with design? We can say that sex-appeal must slowly shift from organic to inorganic, from real bodies to synthetic ones – often just as disquieting – but the process doesn't change: a place's name helps fix the object's image, guiding the feelings the observer or user must feel, to place it in this or that box of imaginary space. By observing the tie between objects and their names, we have the chance to enter into a doubly vital code, needed by the author to finish the project and for the observer to fully understand it.

Things and Words

To talk of names of objects, or of names of places applied to designs, means moving on soft and uneven terrain, held between ineffable boundaries and the profound heart of modernity. In fact, it isn't only about thoroughly seeking ties

between an object and a place, as occasionally happens, or about underlining the evocative traits and dreamy "images" held in the name of this or that specific site, or about losing oneself in linguistic lucubrations that are certainly not needed now.
It is about recognizing the delicate link between words and things, and in particular what George Steiner called "the breach of the pact between words and things" that marks *in primis* modernity. Still, industrial design is the most "modern" thing there is, much more than art or architecture, and infinitely more than "applied arts." Through names – and we mean proper names – design weaves its net of links with space and reality. It invents its *genius loci*, exceeding those atypical and acontextual features that constitute the profound limit and unequivocal code of modernity.
The other pole of the argument – the idea of place – instead seems in some ways more familiar and accessible, helping us reduce problems, restrict terrain to a field and a more noted system of values.

In Principle it was "Barcelona"

The first images that unavoidably come to mind are the Mies chairs for the Tugendhat House and the Barcelona Pavillion. Both have the succinct name of the place for which they were designed. They belong to the rationalist avant-garde era – particularly those made for the German Pavilion of the International Exposition of 1929 – and rise to a fame beyond any expectation.

The *Catalan* armchair, in fact, draws unexpected significance from the crossing of its shape and its positioning; it transcends its own simple modern design and becomes a symbol of perfect synthesis between rationalistic and Teutonic modernity, and the Mediterranean spirit that Mies intended to instill in the design of the building that houses it. We are still in the heroic and progressive era of modernism; the "name" detaches itself from the object in this case but follows it closely; it enlarges and corrects its importance, forming an immediate reference for those who must understand the project's shape and structure.
The chair doesn't really make any explicit reference to the Mediterranean: it has a simple and functional design that completes itself with the meaning of its name; it introjects it, fixes it in an imprecise image that is simultaneously unforgettable. From that moment on, in whatever atmosphere it is inserted, it will manifest the strength of modern thought and its "classical" roots in the forms and places of the Mediterranean.
Barcelona is one of the first cases where design turns to the idea of "baptizing" an object, and it may be the first explicit case where a place name, and in particular a city, is invoked.
Not all "masters" follow Mies in his "metaphysics of the name": he had already designed furnishings for the Weissenhof without giving them a specific name; Le Corbusier continued to prefer "common" nouns and the manufacturing codes of the objects he designed (*la chaise longue...*), and considered them in terms of "equipment" or furniture types, rather than unique pieces; and Aalto

does the same for his glass and curved wood objects, intending above all to go back to the quality of a deeply-rooted artisan tradition.
After Mies, at least according to the *1000 chairs* catalogue by Charlotte and Peter Fiell, Italian designers and architects are the first to follow the German master on the path of proper names. Gabriele Mucchi's *Genni*, Piero Bottoni's *Lira*, Terragni's *Follia* and, abroad, Rietveld's *Zig-Zag chair* relate a simple modern use of the identifying name, still consistent with the look and location of the object.

Locus Solus

In the first years of the postwar period, things didn't change a lot: Aalto continued to call his chair *Armchair* and Eames called his *La Chaise*. And even Mollino cleanly designed *Poltrone per casa Minola* (Armchairs for the Minola home) or *Sedie per Lisa e Gio Ponti* (Chairs for Lisa and Gio Ponti). To find a name dedicated to a place you have to wait until 1952 and the sofa *New York* by Katavolos, Littel & Kelly, where the use of the city name serves to show mainly the aim of conversing with Mies' *Barcelona*, inserting itself into the master line of modernist tradition. Maybe only with Achille and Pier Giacomo Castiglioni's *San Luca* does the relationship with the place name begin to get complicated, acknowledging formal relations and direct suggestions, always under the banner of attempting to juxtapose a solid *genius loci* with a first timid "urban critique" of modernist atopy.

Such criticism resolutely explodes during the 1970s, when design more decidedly nears the avant-garde of Pop art, radical architecture and the corrosive and surprising use of the ironic range. The works of Gaetano Pesce, Archizoom and Superstudio come into play during a cultural phase dominated by the central position of the linguistic-structuralist approach, hence more inclined to play with the object and the meaning of the word that names it rather than use the directly evocative identifying name or to pay tribute to this or that place. During more subtle times, as with the case of Aulenti's *Locus Solus* series for Poltronova, the choice of name became a crucial theme in conceiving the object, interweaving literary ties and revealing the creative process.
Citing the name of a series of garden chairs is decidedly intellectual: the reference is to Raymond Roussel's novel of the same name in which the solitary place is that of the mind and its intrigues. Cerebral complexity is evoked in the garden of a scientist, equally brilliant and original, who has surrounded it with machines. The premise reflects the lexical structure of the author's writing and the physical elements of the object designed by Aulenti. However, Roussel, together with Quenau, Perec and Calvino, is one of the wizards of literary combinatorics, who has charmed such readers and exceptional critics as Michel Foucault.

The Memphis Era

It wasn't easy to give objects place names in the triumphant era of Pop and Conceptual art.

Back when we liked "thinking big" it was easier to find enormous capital letters, tufts of grass from a gigantic lawn, books to sit upon while reading books, or lips to lay down upon and dream of Marilyn. There is obviously a group of furniture and objects – *Atollo, Faraglioni, Duna, I Sassi* – that take their names from a geographic typology. In this case the effect of the geographic object on the physical one is decisive in terms of shape. The designer's process seems to be that of identifying a dominant trait of the geographic prototype – for example the sinuosity of the dunes, or the primitive isolation of the atoll – that constitutes the core of the object's formal reality, which will attract a wider recognition for this trait through the exchange between name and shape. But we need to wait for Gaetano Pesce and his sofa *Tramonto a New York* (sunset in New York) from 1980 – almost the Memphis era – to find a more "critical" use of the name and a readymade reference to a place. Perhaps it is precisely the comparison between two objects with names (nearly) the same that helped bring about changes in the relationship between reality and the word from the 1950s to the 1980s, a explosive period for Italian design. For the three authors of the sofa cited above – Katavolos, Littel & Kelly – New York was the place of modernity and Mies' works, the land of the Seagram building and solid modernist optimism. The name, more than evoking the American metropolis, then served to declare its attachment to a trend, the need to belong to a heroic line of architecture and modern design. In Pesce's work (who also does a very beautiful *Green Street*) New York is an image and a game of skill and irony. The piece of furniture is like a snapshot that depicts, in a man-

ner totally untethered to its function and its rapport with historical trends, the "shape" of a moment in the life of Manhattan. The name, evidently, is like the title of a photo or of a figurative picture. We have, in fact, arrived at what some call "postmodernism," an architectural fact that looks to the past with neo-figurative excitation, nurtured by anti-modern hatred and the desire to quickly bury conceptual severity. In this phase, explains Lyotard in *The Postmodern Condition*, "to speak is to fight."
The name is therefore a clue to enter well-armed in a linguistic game that regulates the relationship between the designated object and the cultural panorama that receives it. The designers of Memphis – the name of a place meaningfully used by an entire company – sensed this very well and from 1981 they conceived of design as "an explosion of colors, an unusual gathering of shapes and materials, a mixture of laminated plastics, glass, marble, metal and wood, furniture as sculpture, but expressive like personalities..."
Even the Memphis designers intended to fight, and the name of the place – Memphis – serves to make everyone understand that the inevitable conception of modernity doesn't necessarily have to limit itself to a nostalgic vision of the historical past while opening itself to the complexity of the present: to neon and pastel lights of American cities, to "bad taste," to a jump in scale, to an imitative pastiche of banal objects and masterpieces from the past.
Then came Aldo Rossi, the opposite of a designer, who designed a series of objects during the 1980s that made clear the manner in which the architectural culture was to respond to industrial design "provocations." *La Cupola, la*

Cabina, and the *Milano* chair serve to specify the limits of design as seen by architects: on one side the city and the history of architecture, on the other side the resistance of banal objects, with their roots in life and above all in people's looks.

Ravello

It is well known that history doesn't have a linear development. We also know the various ways of creating links between objects and place names, which aren't reducible to a sequence, to a progressive line that goes from a narrative use to one that is ever more critical and sophisticated, sacrificed for the ultimate (and very postmodern) separation between form and content.
The impression is that in international design, and especially Italy, various approaches to designing a piece of furniture and object continuously intertwine and overlap, coexisting without excessive conflicts, collaborating in conquering the market rather than daring each other for cultural domination, as is the preference of architects.
The indestructible weight of Citterio's *City* (Is it a place? Which place is it?) converses well with the modern perfection of Magistretti's *Silver*, with De Lucchi's precise eclecticism and with the agility of Sottsass' and Castiglioni's works. Hence, *Barcelona* is still current, demonstrated during recent months by the vicissitudes of the relationship elapsing between Oscar Niemeyer and the city of Ravello. Invited (at the tender age of 95) to design a concert hall for this

small and beautiful center on the Amalfi coast, Niemeyer, who really doesn't travel willingly any more, undertook the long trip and regularly presented his project for the beautiful and interesting Concert Hall, which at the moment obviously has little possibility of being built.

For the Concert Hall, the Brazilian master had also designed a chair that has been realized and put into production by Frau, who is already marketing it. As with Mies' *Barcelona*, Niemeyer returned to the simple and direct habits of the pioneers of modernity and named his chair after the name of the place for which it had been designed: *Ravello*.

As if by magic language and the use of a name, it has thus returned to its primary and narrative function, refusing irony, hypercritical meanings or deconstructive devices, and fixing the image of the chair to the memory of a place, still memorable and Mediterranean.

For us, two further considerations arise from this event, as a possible and temporary conclusion to this discussion. The first has to do with the ambiguous nature and elusive "end of modernism" that seems to have actually left resistant and sensitive traces, helpful paths in both design and architecture, for those who still want to look into the relationship of shape, structure, technology and meaning. The second has to do with the fortunate opportunity, shown once more in the Niemeyer adventure, that Italian design still has to experiment, risk and consolidate its research through a continuous and complex journey, pursued by courage and (often) supported with determination by critics and industrialists. The same can't be said of Italian architecture, stuck

for thirty years in the position of a building industry (to speak only of this) rather than promoting research, experimentation, new materials, or formal experiments that no one wants to speak or hear.
Therefore it's not unusual for Oscar Niemeyer's Concert Hall project (Italian) to remain on paper while his chair, beautiful or ugly, already occupies the display cases of a few scintillating showrooms.

The Theatre of Things:
The Look and Function of Italian Design Objects that Have Names of Places

Silvana Annicchiarico
Curator of the Permanent Collection of Italian Design of the Triennale di Milano

Space seems to be more subdued or more inoffensive than time: you meet people everywhere with a watch, and only very rarely people with a compass.
One always needs to know the time,
but no one asks himself where he is.
One believes one knows: if one is at home, in the office, in the subway, in the street.
It is obvious, certainly – but what isn't obvious?

(Georges Perec)

To really see things, to know them in depth, you occasionally need to separate yourself from their *obviousness*: forget their use, the function normally given them, the point of view and the perspective with which you usually observe them everyday. You don't have to submit useful objects to a metaphysical interrogation, probing and rummaging beyond the *obvious*, so much as to invest the space and the objects that fill it with a glance able to dissolve established codes, consolidated stereotypes, perceptive laziness or classificatory dogmas. As Georges Perec has written, "Space is an uncertainty: I have to constantly single it out, design it. It is never mine, it is never given to me, I have to conquer it."[1] The history of design is filled with investigations of this kind: a project's originality and innovativeness often springs from the ability to insert a doubt into a normally fossilized place, exposing the obvious with a question or impromp-

tu inquiry, to single out a problem where it seemed not to exist and to try to create an answer that makes sense and has value. To do this, it is often necessary to initially perform a sort of *alienation*: to give the object a "different" look, to imagine it outside the usual context, subtracting it from its commodified nature to retrieve its symbolic strength.
A similar alienation is *obvious* (that is, to still follow Perec's tracks, metaphorically and gnostically) in certain proposals from anti-design, or in certain radical scale reversals that designers periodically (and provocatively) devise: one thinks of the meter-high pens and pencils seen in some Milan Fiera exhibitions during the 1930s, or of some gigantic pens of the 1970s, or the objects from Aldo Rossi's *Teatro Domestico* for "The Domestic Project" show at the Triennale di Milano in 1985. As G. K. Koenig writes, "Magnifying a useful object's scale, annihilating its real function (who would put a half-meter-long pen in his pocket?) is a direct way of *alienating* the object, tearing it from the consumer circuit and consequently from aesthetic consumption."[2] This is what designers like Alessandro Mendini and Gaetano Pesce have often done, designing and making very large objects in heavy and durable materials, almost as if wanting to *monumentalize* the object itself, tearing it from its practical function and repositioning it above all as a "self-representative" artifact (for example, one thinks of the *Moloch* lamp of 1971 by Gaetano Pesce, a five-meter-long out-of-scale reproduction of a *Naska loris* office lamp). The same Mendini writes about similar operations: "It is about practically useless things for man and for the system, *second-rate* objects that demolish the usual logic attributed to their func-

tion with their extravagant presence. Their goal is to serve man by meditating on his own actions on futile objects, and on a possible different organization of his livable surroundings, intended as a direct extension of inner life."[3]

Even the objects grouped in this show carry out an analogously alienating effect. They also perform a *scale reversal*: not through dimensions or enlargement tactics but through a *nominative strategy*. Their *theatricality* (representative strength) derives rather from their name. Name-world, name-space, name-architecture. When a sofa is named for a city (Francesco Binfarè's *Tangeri*), a chair for a state (Antonio Citterio's *Panama*) or a table for an architectural standard (Mario Bellini's *Colonnato*), you can't pretend that nothing has happened, and continue to believe that the chair is only a chair, or that the sofa is only a sofa. Who can do this, risking the same error (with the same *dismissal*) as Hamlet's mother? "Do you see nothing there?" asks Hamlet of his mother. "Nothing at all, yet all that I see," responds the Queen. That nothing was *no less than* the specter of the dead King, his father and her late husband.[4] The *specter* of these objects (the aura they emanate, the meaning to which they refer) is, in fact, the *specter* of the world. It is the large-scale raised in the small. If anti-design enlarged objects to create a world from them, these objects chase the world to make it smaller (but also symbolize it, conserve it, invest it) in an object, so as not to lose its memory, to remind us that even the home is a world, ours, and that every *interior* devotes itself only in as much as it is a simulacrum of a coherent and complete world.

Georges Perec once more reminds us of it: one begins, like Joyce's Stephen

Dedalus, by naming ourselves and our own address, arriving at the world's nomenclature and universal geographic space ("Georges Perec, 18 rue de l'Assomption, Stairs A, 3rd Floor, Right-hand door, Paris 16°, Seine, France, Europe, World, Universe").[5] These objects make the same journey, but they do it with an ellipsis, with a nominative somersault; they are already of the world and they make it present for us. Within them, in their "being there," space-function-language becomes an inseparable but discontinuous ensemble due to the jump in scale imposed by the name, or by the theatricality that the name superimposes on the function. A lamp called *Manhattan* or a chair called *Broadway* transmits a discontinuity of the domestic space that imposes on the inhabitant a continuous re-examination (or an incessant reflection) of the same space, producing a discontinuity in language that pushes us to re-examine the procedures for obtaining meanings in that language.

Certainly: the city of Milan and Aldo Rossi's chair of the same name are not equal, nor are Stefano Giovannoni's *Rimini* cutlery drainer and the Romagnolo seaside town. And yet, these names open cracks, insinuating a metonymy; they push the chair as well as the cutlery drainer to "act" another part, to impersonate a different role. Or to represent themselves in a non-accidental net of echoes, cross-references and relationships. "Every place," wrote Aldo Rossi, "is certainly unusual in the measure in which it possesses endless affinities or analogies to other places."[6] These objects carry a real internal desire for topological otherness, of enormous relational nature, of analogical allusiveness. A bit like the closets Aldo Rossi had wanted to call *Le cabine dell'Elba*, not only in homage to the *type* that the project had suggested or as a token of the place where the project itself had taken shape, but also and above all as a revelation of a far-sighted will that intends to make a "small house" from a piece of furniture, and to give it that shape, or – vice versa – that intends to make an *exterior* structural model the matrix upon which to reshape even the *interior* spaces of a home. "I called them cabins because they are effectively referred to and used in that way, but also because it seems to me to be a minimum size for living, an impression of the summer."[7]

Here is really the core of the problem: Italian design objects that carry the names of places (of a city or state, of a morphological landscape element or of a typological architectural feature) perform a triple semantic shift or, if you prefer, a triple jump in scale: between *small* and *big*, between *interior* and *exterior*, between *function* and *appearance*. The three planes are obviously intertwined and correlated between themselves. At the first level, that of *dimension*, it is obvious how a small object with the name of a big city tends to inevitably take on a representative function that conditions and widens its most obvious and immediate functional characteristics. At the second level, that of *position*, the name of a city or of a place bestowed upon an object for a house's interior makes it so that the same object introjects the world's *external quality* inside domestic walls and in some way amplifies and dilates it. Finally, at the third level, that of the *theatrical*, it is obvious how a piece of furniture named *Casablanca* or a vase called *Yemen*, both designed by Ettore Sottsass, give off a representative power that alters and inhibits a merely functional enjoyment of the object, raising an entire series of symbolic, highly imaginative or narrative bonds. Through the presence of objects like these, in short, everyday domestic space defers to other living spaces or even to those only dreamed of, framing those other imaginary places a bit like illustrations or pictures hanging on walls, or like a television monitor. With the difference being that pictures and television programs show the world in images, while the objects in this show conjure the world *in absentia,* leading us to imagining it rather than seeing it iconically. If Aldo Rossi gives his domestic utensils the shape of architecture and calls a coffee maker *La Cupola*, and if many projects by Enzo Mari carry the name of an island, that means that the end objective of these objects is to set themselves up as microcosms holding traces and marks of something transcending them, ultimately demolishing the conventional divisions between *far and near*, between *here and there*, between *public and private*, between *I and world*. In fact, the objects in this show contribute to making the world a *cockpit*, in the sense proposed by Bruno Munari: "Cockpit is the environment / adaptable to the inhabitant's personality / transformable in every moment."[8] The world returns to being close at hand, it redesigns itself as an extension of the body; it becomes tangible and easily manipulated once more. The operation is totally metaphorical, but not useless or superfluous, nor relegated to a mere signifier for nostalgia. If these objects conjure up the world (as the objects with animal names gathered in the *Animal House* show conjure up nature), at the same time they force us to reflect on our relationship with manners, shapes and our sense of *representation*. "Elephants," wrote Jonathan Swift, "are generally drawn smaller than they really are, but a flea is always drawn bigger." The objects exhibited in this show stage an analogous representational paradox: they don't alter the proportions of things, but make it possible for the world to be held in a room, without making the world necessarily smaller. On the contrary, it is left beside us. So as to have the world in the home. As if transferring, also on a symbolic plane, the miniaturization process that has for some time invested useful objects with strong technological connotations.[9] Design objects named after places convey on the world (based on the idea we have, the relationship we maintain with its shapes, spaces and geographies) a sort of emotional *reduction*, an emotional closeness, a subrogating continuity. Is it a way to exorcise the mixed-up de-territorialization of the globalized community, covering the only space that is still definitely *ours* (the house, the domestic interior) with an allusive system of signs made up of urban nostalgia, geographic indications and architectural morphologies with a very prominent identity? But not only this. It is, in fact, thanks to objects like these, once considered inseparable in their nexus from shape/function/language, that, as Fulvio Carmagnola has written: "The territory of design broadens, it invades the dimension of figurative art and architecture. Within the relative intransitivity, in the ironic game of shape that protrudes on function, that covers itself with exceeding uselessness, it shows itself to be a different cultural value, that implicates a complexity greater than enjoyment."[10] When a product expands itself to the scale of a city, it certainly doesn't suddenly become architecture, but it evokes the idea in itself. With a twofold result: from one side "the perceptive figure of the object becomes uncertain and fluctuates between different levels," and on the other side the same idea of the city "transforms itself into an agglomerate of objects beyond scale (...), a catalogue of knick-knacks, an attic of dreams."[11]

Also in this case, as with the two previous exhibitions from the "Words and Things" series (*I'm No Lady* and *Animal House*), preliminary research has allowed for singling out and mapping a very conspicuous group of Italian design objects with place names (several hundred, more than in the two groups from the two preceding shows). Aside from a few clamorous cases of geographic names chosen as a precise company marketing strategy (the Danese catalogue has plenty of products with island names, Memphis has chosen to name numerous objects after exotic places or big international hotels), this time the group is less homogeneous as those proposed in the preceding shows, or to say the least, it presents a greater variety and number of figures: it goes from objects that take on the architectural type or model name which they often resemble (the coffee machine *La Cupola*, the table *Colonnato*, the chair *Capitello*) to those that adopt and actually have the name of a city, indiscriminately close (the *Milano* chair, the *Sanremo* lamp) or faraway (the small table *Kyoto*, the *Tangeri* sofa), as they gradually become objects that perform nominalistic scale enlargements evoking the character of a morphological landscape or natural environment (the *Atollo* lamp, the *Lido* sofa) or even taking on an exotic island name (the *Bali* lamp, the *Lipari* table) or the name of a state (the *Yemen* vase, the *Louisiana* armchair). The variety, as well noted, is everything but gratuitous and responds to a precise logic: it goes from small to big, from type to city, from morphological element to the island or state. Almost as if wanting to embrace the *external quality* in its *integrity* and *plurality*, the job of geo-architectural naming unravels according to an inclusive tactic that doesn't ignore a single aspect of the world; everything harnesses it and ties it to a thick net of relationships. There is a conspicuous subset that pushes itself even further. It is the one made up of names that call a delivery van a river (*Lambro* by Innocenti) or a car or a motorcycle a state (the *Nevada* and *Florida* motorcycles by Guzzi, Ferrari's *America* and *California* models), almost wanting to send the world traveling: transportation vehicles used to move from one place to another become in themselves places, mobile spaces, mutant vehicles, not only a "means" but – precisely – "states" and "modes" of an existential condition based on mobility and, once again, on the dream of the world's total attainability and availability. In short, enclosed in a room or made to travel, the names of these objects follow the same goal: that of stripping the world of its otherness; making it familiar, domestic and attainable. The adopted classification criteria to give the show's objects a legible structure couldn't but mimic the variety and inclusiveness of the types of names. The show's sections consequently describe and mirror the classifications written in the name of the objects, concentrating them – as usual – in a single room. To repeat, the coexistence and the cohabitation of the objects in a single narrow space sends them back to a very different scale, making a discontinuous but coherent, unified and potentially infinite *expanse* of the space itself.

As in the entire "Words and Things" series, avoided in this case is an abstract and idealistic nominalism, proposing instead a more articulated and complex vision that tries to project and reconsider shape, function and the names of things within the delineated scenario of social and emotional use. The aim is to trace a kind of logistic intuition, or a phenomenology of reception of design objects, convinced that, as shown by Canadian scholar Jan Hacking, certain classifications, when they are assimilated by people or by their environment, modify in turn the way in which people look at themselves and the objects in their world.[12] Hacking states, "Names are part of an immense world of customs, institutions, authorities, connotations, stories, analogies, memories, ghosts. (...) Naming engages places, certain sites, and happens in precise moments. But how can a name that begins its work of creation need authority? It needs to be put into service within institutions. A name takes on its function only when a social history is itself in action."[13] Even in the case of the combined exhibitions in the "Words and Things" series, it is the social history, in the end, that gives meaning and value to these objects as far as their names are concerned. And this is what makes sure that they really serve to make the world more habitable and comfortable, and to produce here and there some illusive mental mirage.

Notes

1. G. Perec, *Espèces d'espaces*, Editions Galilée, 1974, p. 122; It. trans., *Specie di spazi*, Bollati Boringhieri Editore, Turin 1989, p. 110.
2. G. K. Koenig, *Wasps and Elephants*, "Ottagono" n. 48, March 1978, p. 65.
3. From a reading by A. Mendini, "Ottagono" n. 39, December 1975.
4. W. Shakespeare, *Hamlet*, Act III, Scene IV.
5. G. Perec, op. cit., p. 113.
6. A. Rossi, *Autobiografia scientifica*, Pratiche Editrice, Milan 1999, p. 54.
7. ivi, p. 56.
8. B. Munari, *Da cosa nasce cosa*, Laterza, 1981, p. 196.
9. J. Baudrillard, *Le système des objets*, Gallimard, Paris 1968, p. 72 and followings.
10. F. Carmagnola, *Della mente e dei sensi. Oggetti dell'arte e oggetti del design nella cultura contemporanea*, Edizioni Anabasi, Milan 1994, p. 152.
11. Ivi, p. 151-2.
12. J. Hacking, *Il caso domato*, Il Saggiatore, Milan 2002.
13. M. D'Eramo, *Quei nomi che riplasmano l'anima*, Interview of J. Hacking, "il manifesto," April 26, 2002.

Design by Aldo Rossi, 1984

Objects/Architecture

Archetypes, prototypes, canonical models from architectural history: towers, colonnades, cupolas, capitals, and skyscrapers. The objects included in this section of the exhibition are those with great coherence and clarity: they play on scale reversals and transfer to narrow *intérieur* domestic spaces typologies and stylistic elements conceived for a grand scale, as much in the urban dimension as in the architectural-monumental one. This world of familiar objects, miniaturized and contained in a room, resonate with archaic echoes but also of impossible utopias: Massimo

Morozzi's *Babele* étagère conjures up the Bible and ancient Mesopotamian *ziggurats*, Mario Cananzi and Roberto Semprini's *Tatlin* sofa brings Constructivism and the impossible dream of the Russian revolutionary architect to the bourgeois living-room, while Aldo Rossi's *La Cupola* miraculously succeeds in making a miniature monument of a coffeemaker. Small and large, interior and exterior, present and past: in the game of transfer and analogies, of semantic shifting and stylistic transitions, the thought that is at the base of similar projects succeeds in problematically reopening the question – never completely resolved – of the methodological and conceptual relationships between design and architecture, proposing a third direction – metalinguistic and conscious – between the dogma of proclaimed absolute unity by certain rationalist trends and the postulate of a total and unrelated autonomy sustained with excessive *nonchalance* by certain contemporary trends.

Trumeau

Sideboard
Piero Fornasetti and Gio Ponti
1951 – 1970
Piero Fornasetti
1980
Immaginazione srl

Piero Fornasetti puts into effect extravagant architectural decorations on the functionalist shapes of Gio Ponti's sideboard from 1950, covering all surfaces with adaptations of seventeenth- and eighteenth-century prints. Reproduced on the front of the cabinet are the classic courtyard of Milan's Accademia di Brera, the University of Genoa's stairs and the façade of a Milanese palace by Bramante. A rustic stone wall texture is rendered on the sides. Inside the cabinet doors a palace interior is depicted in deep perspective with trompe l'œil technique. Here the decoration is not a simple addition to a given scheme, but expands the volume of the piece of furniture, giving it life, depth and historical richness; it employs a composition that absorbs large-scale architecture within the smaller scale of a piece of furniture, not only by suggestively superimposing illusion, but also through theoretical implications and relationships both mnemonic and emotional.

Triennale di Milano Collection

Pirellone

Lamp
Gio Ponti
1967
Fontana Arte

About ten years after the realization of Milan's Pirelli skyscraper (1956) – a functionalist symbol of a city's desire "to look to the future" – Gio Ponti recaptured the formality of his most famous building in a series of lamps for Fontana Arte. Created in varying heights, they are not only reminiscent of the skyscraper's slightly ogival shape, but the lighting elements are distributed at different levels, suggesting the floors of a building built along a vertical axis. The name *Pirellone*, given to the tallest lamp (the shortest is called *Pirellina*), is allusively ironic yet affectionately evocative, almost wanting to evoke the structural model's formal perfection. Ponti dematerializes the imposing and massive skyscraper by transforming its formality into an airy and light structure where the geometric distribution of light allows a glimpse of the design's internal structure.

Fontana Arte Collection

Capitello

Armchair
Studio 65
1970
Gufram

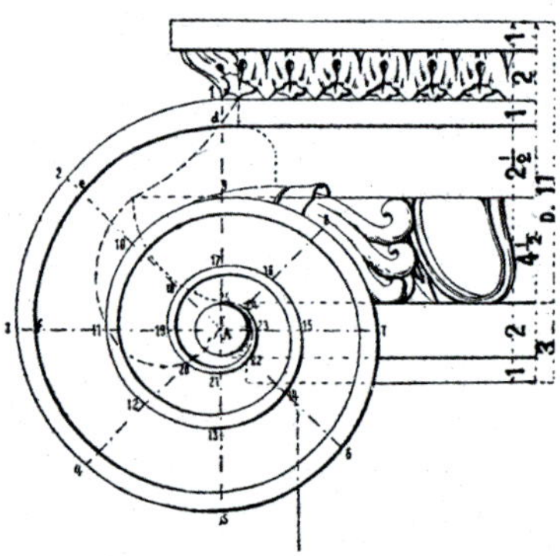

Made of lacquered expanded polyurethane, this armchair, designed in the early 1970s, imitates an Ionic column with the dimensions deliberately exaggerated. The design's attitude is more ironic than archeological. As in the entire *Attica* series, classical revival enlists an anti-design idea that precludes postmodern humor through a twist of displacement: the architectural element classically used to support the temple's large mass is reduced to a formal structure destined to hold the body within the domestic living room. Created as an assault on good taste, it has become a furniture icon.

Studio 65 Collection

Il Colonnato

Table
Mario Bellini
from 1977 to 1982
Cassina

Far from the postmodern aesthetic of appropriation, Mario Bellini lavished his classical, renaissance and neo-classical taste for beauty and history in this design for Cassini, attempting to give each separate shape in the table's design an aesthetic dimension. Applying a monumental form such as a column to a piece of furniture for a domestic interior, Bellini continues to question the relationship between design and architecture, supplying a far-from-dogmatic answer that makes something new out of the historical and formal past. Made out of Roman travertine, Persian travertine or black marble, and held by 3, 4 or 5 marble columns, the *Colonnato* (colonade) table re-proposes utopia with a design that knows not to forget architecture (as confirmed by the series of other tables planned by Bellini for Cassina: *La Basilica, La Rotonda, Il Castello, La Corte* and *La Loggia*) even without trying to restore the unifying concept of design that, according to Bellini, would be typical of Bauhaus or the Ulm school.

Cassina Collection

"Each design project by Bellini has a small piece of history, where history is not so much a quotation as a filtered and transferred heritage – also ironic – in contemporary work, with everything always controlled and measured within the constraints of the industrial process."
(Aldo Colonetti, "The Language of Mario Bellini," in *Mario Bellini: A Journey Through Architecture, Furniture and Machines*, Galleria Civica di Arte Contemporanea di Trento, 2000, p. 7).

Le Cabine dell'Elba

Closet
Aldo Rossi
1980
Molteni
1983
Bruno Longoni

For Aldo Rossi, the cabin represents the most basic and universal architectural shape: the "small house" is the essence of architectural gesture and fact. These cabins (those of Elba mainly) are the origin of Rossi's different designs and become a constant compositional element (house-cabins fit next to each other around a square with a central building in the Chieti student housing project of 1973; they were an even more important new element in the 1977 plan for Florence's Administrative Center; they were key in generating the "Teatrino Scientifico" in 1978; they were suspended on stilts in the Klösterliarcal project in Berne in 1982; and for the restructuring project of the Campo di Marte on Venice's Giudecca in 1985, the house-cabins represented a "square in the traditional English urban sense"). With the Closet-Cabin of 1980 for Molteni (produced later by Bruno Longoni), Rossi brings his favorite model within domestic walls, consistent with his vision of an inner space built like the exterior, with architectural languages and forms from the city. Transformed into a closet, the cabin returns to its original function and, indoors, marks the end of summer, preserving a joyous or sad memory of the beaches from which it came.

"In reality, I was simply aware of the particular and universal character of the cabins on the beach. Not only Elba's...I saw thousands of cabins, from the beaches of the Mediterranean, to Californian, to Argentina... The cabin is one small thing: the reduction of the house, the idea of the house."
(Aldo Rossi, *Le cabine dell'Elba e altri mobili*, Bruno Longoni, Atelier d'arredamento, Milan 1993, p. 3).

"There are instead these small houses with no location, since the location is within, or one identifies oneself with those who inhabit them for a brief period, a period impossible to calculate.
The cabins without exception have four walls and a gable; and the gable has something that isn't only functional, as it suggests a flag and also the color. The striped color is part of the architecture, recognizable, perhaps the most obviously architectural part."
(Aldo Rossi, *Autobiografia scientifica*, Nuova Pratiche Editrice, Milan 1999, p. 57).

Triennale di Milano Collection

Ponte

Small table
Angelo Cortesi
1981
Fiam Italia

Ponte (bridge), a single shelf in curved crystal, able to simultaneously hold objects and be penetrated by light. Inspired by the prototype for a small curved crystal stand realized back in 1934 by Pietro Chiesa for Fontana Arte, Angelo Cortesi uses the glass technology of Vittorio Livi's Fiam company to design an object that liberates glass from its historical role as an ornamental complement, making it into furniture. Thanks to a high-precision technology, Fiam is able to render glass into curved shapes, modeling and cutting it to the specifications of the designer. The end product has surprising strength, purity and transparency.

The nominal jump to the morphology of an architectural prototype like the bridge is almost compulsory: both for the structure's shape that, as a work of engineering, challenges and suspends gravity, as well as for its obvious ambition to suggest – both in terms of content and from a technological point of view – a *bridge* between present and future.

Fiam Italia Collection

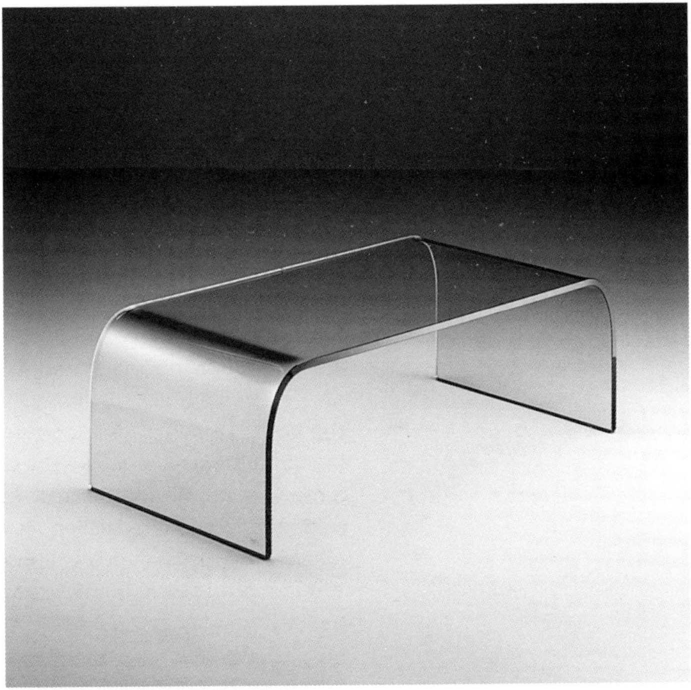

Babele

Shelving unit
Massimo Morozzi
1987
Fiam Italia

The name conjures up the Tower *archetype* par excellence: the Assyro-Babylonian *ziggurat* has always embodied the "artificial symbol of Olympic heights, where men could approach the gods or invoke their descent." (M. Scolari, "Forme e rappresentazione della Torre di Babele," *Rassegna*, n. 16, December 1983, p. 5).
The favored historical representation of the Tower has been the Renaissance cone shape, attested to in many sixteenth- and seventeenth-century European paintings, after the parallel towers of the miniature Medieval codexes and before the Assyriologists reconstructed the square shape with archeological rigor in the 1800s and 1900s.
Massimo Morozzi's *Babele* (Babel) bookcase, with its emphasis on trapezoidal curves, retrieves and re-purposes the turreted and soaring structure. The reference to the ancient Mesopotamian myth of the Tower is metaphorically heightened by the apparent frailty of the material and by the function that the object accomplishes: other than holding knick-knacks, the unit is perfect for holding books, conjuring up the *confusio linguarum* that the biblical story links to the *impious* project of the Tower's construction.

Fiam Italia Collection

La Cupola

Coffee pot
Aldo Rossi
1988
Alessi

Alessi Collection

Almost a "monument in miniature," the *La Cupola* coffee pot (cast in aluminum, with handles and knobs in black or light blue polyamide) symbolizes the link between architecture and the domestic landscape that Aldo Rossi has pursued and investigated in the past. Described as "a dome held up by a molded cylinder, more or less functional and decorative," the coffee pot underlines the use – even in interior landscapes and the universe of useful objects – of Rossi's utopian notion of the "analogous city," obtained by moving monuments specific to certain cities to new spaces where they don't belong, thus contaminating a virtually unlimited number of cities. A coffee pot-temple that is both tower and basilica, *La Cupola* celebrates the rite of coffee, ironically evoking the sacred and profane. It follows the dream of a domestic landscape constructed and designed to be precisely analogous to urban space.

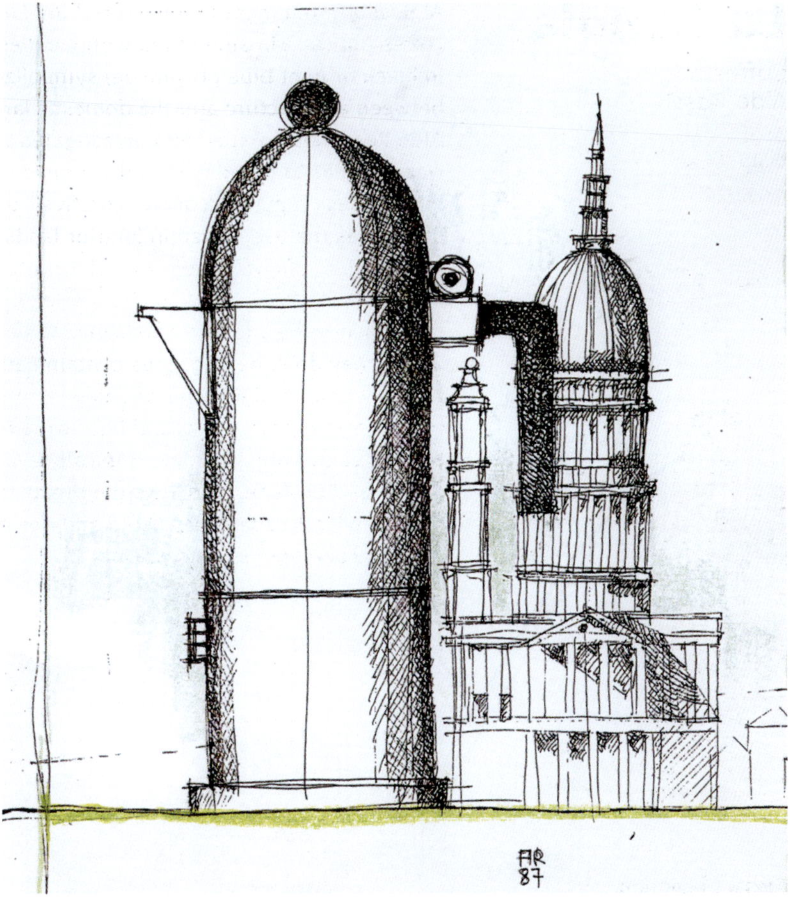
AR
87

Tatlin

Seat
Mario Cananzi
Roberto Semprini
1989
Edra

The irruption of architecture within domestic walls upsets and redefines the traditional proxemics of seating in the intimacy of the bourgeois home. Inspired by Vladimir Tatlin's Constructivist *Monument for the Third International* of 1920, Mario Cananzi and Roberto Semproni's design symbolically borrows the same spiral and helical structure. The reference is not to a canonical typology of standard architectural language, but to an unaccomplished utopian project (the Tatlin tower aspired to exceed the Eiffel Tower in height), making a sort of totem that attacks the philosophy of traditional furnishings: unlike other pieces of upholstered furniture, *Tatlin* refuses to be leaned against a wall and made into a kind of furnishing backdrop. On the contrary, it revolutionizes the living room by claiming center stage, following a design that fuses Constructivism and Romanticism in a surprising *coup de théâtre*.

Edra Collection

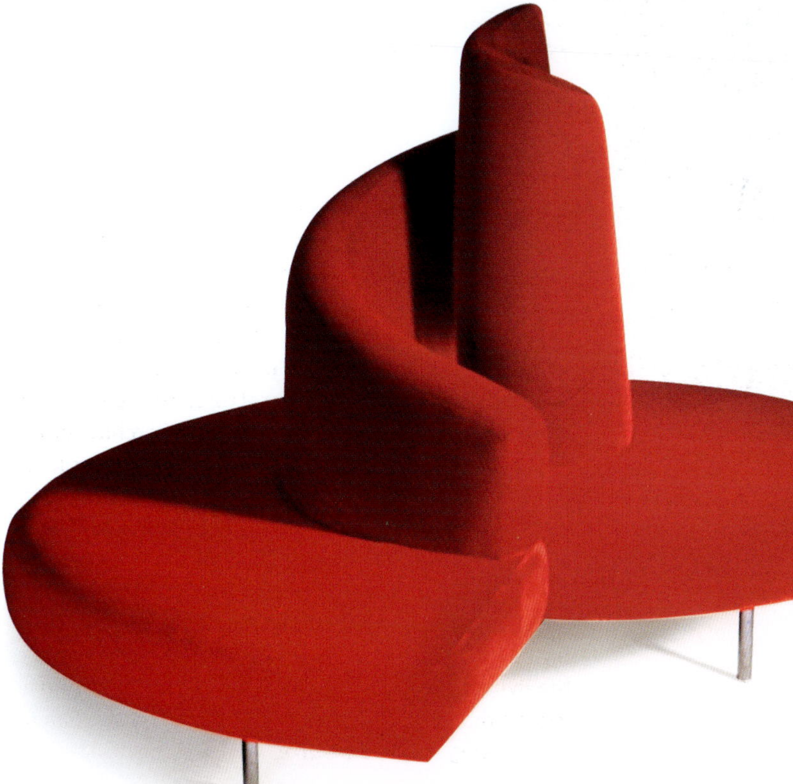

Montjuic

Lamp
Santiago Calatrava
1990
Artemide

Santiago Calatrava confidently works with scale changes in his design process (from jewelry to sculptural scale to large bridges, train stations and airports). He uses similar shapes to design a lamp, a roof or a bridge: manipulating scale, he produces shapes that can be seen and perceived in many ways. *Montjuic* is emblematic: it is both a telecommunications tower (1989-1992) built on a hill with the same name in Barcelona, and it is a lamp for domestic interiors. "The tower is made from an inclined shaft with a metal structure on three supports. Its design relates to the entire construction's resulting vertical weight. The inclined shaft coincides with the sun's rays on the day of the summer solstice in Barcelona, and the tower thus becomes an enormous sundial" (Luca Molinari, *Santiago Calatrava*, Skira, 1998, Milan, p. 52). The lamp, in black expanded polyurethane resin with an opaline white diffuser made from methacrylate, has the same structure as the tower. In addition, if the tower absorbs natural light in performing the function of a timepiece, the lamp is an element that guarantees – on another scale – the control and visibility of space by artificially generating the same light that the tower uses naturally.

Triennale di Milano Collection

Objects/Cities

Incorporating a city into a room. Transferring a mental image onto an object. Preserving the myth, the memory and the memento. Or, more radically, making a concrete and tangible object from a symbolic avatar, representing and giving form to a distant and unknown city that exists in everyone's imagination. Italian design objects with names of cities actually also perform – in addition to the obvious utilitarian functions that are expressed in concrete and specific gestures (sitting, lying down, turning on the light, preparing coffee, and so forth) – an important supple-

Design by Aldo Rossi, 1984

mentary function on a symbolic, emotional and relational level: they make links between the home and the worlds they evoke; they facilitate journeys between here and there, opening up domestic space and making it available to live in the world, or to be occupied by it. Reducing a city to a standardized object ends up, in fact, enlarging the space where the object is destined to be received, following a double semantic movement that effects miniaturization in order to produce its opposite. Amphibological objects are always marked by a double nature, having the capacity to indiscriminately live as much in the functionality of space as in the habitat of metaphor. Italian designs with names of cities have various types of relationships with their point of reference: they can share a few basic values with their reference city (sobriety and elegance in Aldo Rossi's *Milano* chair) or a few specific connotative elements (the performative spectacle of Gaetano Pesce's *Broadway* chair); they can allude to a very personal link in the designer's biography (the *Cobán* coffee maker that Richard Sapper named after his father's birthplace); they can establish themselves as an signifier for ironically critical intentions (Gaetano Pesce's *Tramonto a New York* sofa) or be critically ironic (Archizoom's *Sanremo* lamp). In any case, similarly named objects create a coalescence of memories, echoes, reminiscences and fantasies. And they supply additional signposts, neither pleonastic nor arbitrary, that compliment our need for access – even in a room – to a personal and emotional map of the world.

Chester

Armchair
Renzo Frau
1912
Poltrona Frau

Born in Edwardian England and revisited by Renzo Frau at the beginning of 1910, *Chester* symbolized for the manufacturer not only virtuosic leather workmanship, but also a furniture philosophy that expressed grandeur and pride, solid ties with tradition, middle-class ostentation, comfort, yet also status. The name not only recalls the English city Cheshire, but also the encoded rituals of parlor conversations and teatime, elegant manners and refined decor, aplomb and harmony. Some consider *Chester* to be the archetype for the modern armchair that transformed every living room into a *fumoir* or *boudoir*, offering the ideal backdrop for social self-portraits of the Italian bourgeoisie.

Poltrona Frau Collection

"Even the noted 'sense of restraint' found in the English tradition of the most famous cabinet-makers is further emphasized here; upholstery work is pared down, each form of opulence eliminated, decorations are banished, leaving only the leather craftsmanship to model the piece's sculptural form, accompanied by soft padding that follows the curves with regular and rhythmic pleats, restrained and pillowed by *capitonnée* work, enhanced by borders and gathers." (Daniele Baroni in *Poltrona Frau*, Decio Carugati, Electa, Milan 2000, p. 34).

Carimate

Chair
Vico Magistretti
1959
Cassina spa
2000
De Padova

In 1959, Vico Magistretti designed the Club House for the Carimate golf club in Brianza, about 35 kilometers from Milan. For this important architectural work Magistretti designed specific furnishings (lamps, chairs, etc.) that included *Carimate*. The chair was produced by Cesare Cassina, who understood that it was a universal, timeless object, adaptable to various circumstances and needs, a piece of furniture suited to the most disparate places.
Carimate was born as a refined redesign of an anonymous object: a rustic chair in wood and straw, with a bold note of red lacquer on wood. In this case the place and architectural name are transferred to the object as proof of the origins for which it had been conceived, and as a declaration of its roots.

Triennale di Milano Collection

Sanremo

Lamp
Archizoom
1968
Poltronova

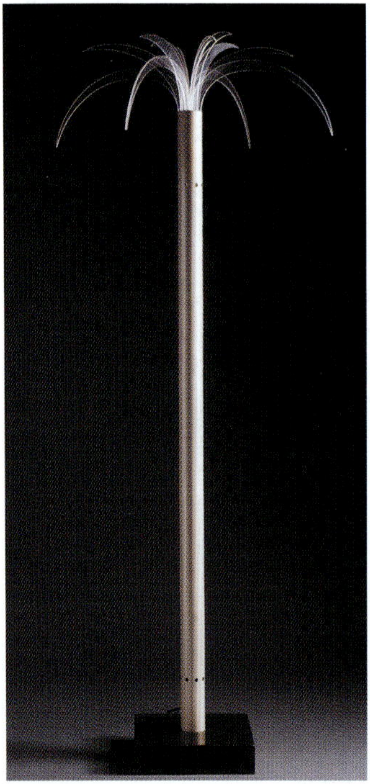

The tourist-postcard symbol of the Western Riviera seaside town (the palm that decorates the seaside promenade) is elevated to a pop signifier in this Archizoom lamp that mocks functionalism and "good design" with an evidently kitsch design gesture. Designed during the rebellious years of the '70s and '80s, it is made from a lacquered metal shaft that holds a bulb at the top, illuminating a palm leave silhouette in Perspex. The *Sanremo* lamp wouldn't be the same if it hadn't been given a name that alludes to an Italian pop cultural fetish (*il Festival della Canzone*, *La Riviera dei Fiori*, the rituals of fake-snob mass tourism) for which we are simultaneously enthusiastic and critical. A typical example of the new décor proposed by Radical Design, *Sanremo* is created for an obviously eccentric and independent place within the home; rather than an element to be added to a homogenous and coherent grouping, it functions as a sign of displacement and as a link to the outside.

Poltronova Collection

Manhattan

Lamp
Nanda Vigo
1972
Arredoluce

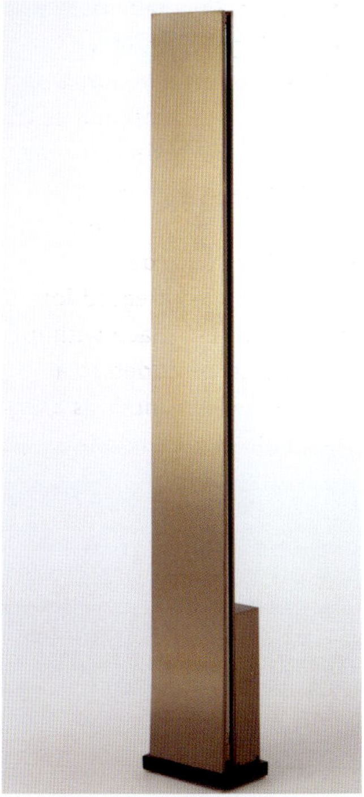

Like an illuminated skyscraper at night: two sandwiched slabs of reflecting material contain – nearly hidden in order to conceal the light's origin – a long florescent tube. The object world of Nando Vigo functions like this: made of artificial light, amplified by reflecting surfaces, glass and mirrors, he creates domestic landscapes through variations in light. With this lamp, designed at the beginning of the 1970s for Arredoluce, the name Manhattan is almost obligatory: not only for the obvious formal appeal of the typical New York skyscraper, but also for the innovative technology that causes the light to change intensity, like the nocturnal phantasmagoric techno-illumination of the Big Apple. Also the materials (aluminum and steel) speak the language of modernity and contribute to making believable the metonymical transition suggested by the object's name: from a skyscraper-lamp to the city that made the skyscraper its symbol and its dream.

Triennale di Milano Collection

Kyoto

Low table
Gianfranco Frattini
1974
Ghianda

Low. Almost resting on the floor. Ideal for furnishing a Japanese interior, a novel by J. Tanizaki or a film by Ozu. In finely inlaid natural beech wood, Gianfranco Frattini's low table avoids the risks of "chinoiserie" thanks to its perfect formal balance and compositionally elegant design that stands on its own while synthesizing a million-year-old tradition into the culture of furniture. The name, here more than ever, *isn't optional*. It's an essential element in the object's reception: not a clever exotic wink, but a precise geographic reference revealing a cultural model.
In Italy during the 1970s, similar objects contributed to the deprovincialization of taste, stimulating comparisons with various dominant North American models.

Triennale di Milano Collection

Boston

Cocktail service
Ettore Sottsass with
Alberto Gozzi
1979
Alessi

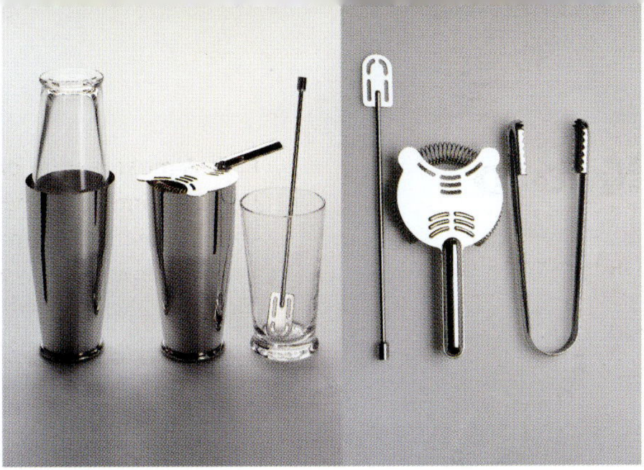

Elegant and refined like New England taste, but also modern and essential in the style of Sotsass, the *Boston* cocktail service (consisting of shaker and bottle cooler, special stirrer, strainer, ice tongs and whisk) emphasizes circular shapes and a material flexibility (stainless steel) to create a group of coordinated objects that are not merely functional.
The shaker, highly professional, is derived from a type that, according to legend, was invented by an unknown bartender aboard a transatlantic liner on route between Boston and Europe at the end of the nineteenth century.
The name alludes to a modernity that has firm roots in tradition. The formal perfection of the individual components, and their integration as a set, recalls the perfect cocktail as prepared by the confident gestures of David Niven in a Hollywood comedy from the 1950s.

Alessi Collection

Tramonto a New York

Sofa
Gaetano Pesce
1980
Cassina

With seats, armrests and back made from polyurethane cushions covered in fabric that reproduces skyscraper facades, and a big red cushion that evokes sunset, holding the structure together, *Tramonto a New York* is an example of an ironic and radical look that Gaetano Pesce projects equally on design and architecture. Here, the scale reversal that interjects the urban skyline within domestic walls has a strong controversial connotation: the "sunset" suggests nothing romantic, and the miniaturization doesn't legitimize any power fantasy, instead indicating evidence of a decadent tendency.

Triennale di Milano Collection

“I have named many of my objects after places where I have lived or worked (think of the *Cannareggio* sofa, or the *Joliette* vase, whose name was taken from rue Joliette in Marseille where there is an extraordinary glass processing center). In the case of *Tramonto a New York* (Sunset in New York), I had initially wanted to call it Tramonto *di* New York (New York’s Sunset) to express my impression of a capital in decline and headed towards the loss of its cultural centrality. Then Cassina transformed that *di* into *a* to give it a less apocalyptic connotation. (From a conversation with Gaetano Pesce, Milan, July 26, 2002)

Casablanca

Sideboard
Ettore Sottsass
1981
Memphis

Triennale di Milano Collection

It is a symbolic piece for the Memphis group, that collective of young creators – architects, designers and graphic artists – founded in Milan in 1981 around the charismatic figure Ettore Sottsass. Anti-functionalism, exaggeration of formal invention and deliberate stylistic incoherence are the particular expressive clues in a design that puts under discussion the most ingrained procedures and stylistic elements in furniture design production. Against modernism's standardized and hyper-rational universe, Memphis proposed furniture with components that were more emotional than functional, more artistic than commercial, assembling usual shapes, materials and bright colors intended to astonish and provoke fantasy. The *Casablanca* sideboard (also ironically defined as a "buffet") is one of the group's seminal creations: Sottsass realized it in laminated plastic, re-evaluating a material previously considered "vulgar." He gave it a totemic shape that desecrates formal values and the domestic context's normal functional parameters with a vaguely anthropomorphic artifact full of echoes and archaic reminiscences. The sideboard's surface is

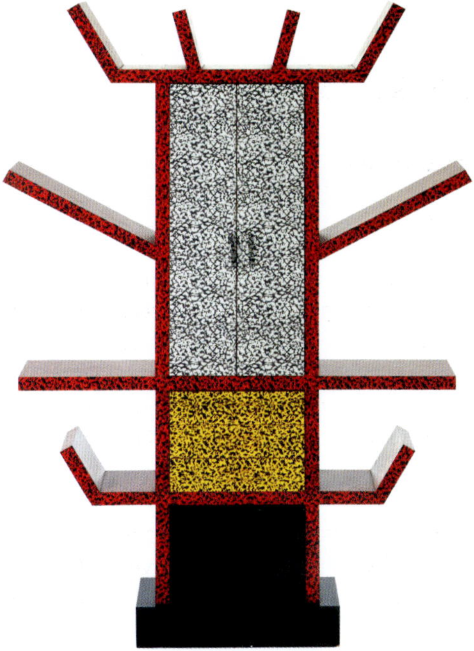

uniformly decorated with a "marbling" of spores and bacteria (the texture, not coincidentally, is called *Bakterio*) destined to spread almost epidemically in graphics and publicity during the '80s. The name confers an emphatic exotic aura on the object and sets it within the mental space that exists between Michael Curtiz's film of the same name and Paul Bowles' flight from celebrities in search of their "tea in the desert."

Milano

Chair
Aldo Rossi
1988
Molteni&C

In this object from the late 1980s, Aldo Rossi reworks the simple elegance of a chair with an antique tradition: in parallel wooden strips, the backrest and seat are typical of Tuscany in the fourteenth century. Rossi, in reality, refers more to a "type" than a "model," and conjures it up again by filtering it through a redesign that produces an up-to-date effect. Even the name is a symptom of such a procedure: not an allusion to the object's geographic-cultural *roots*, but a determined and energetic vindication of the designer's true *milieu* (geographic, but also social and cultural). With the Vulgate imagination of the Lombard capital, Rossi's chair shares, in fact, a simplicity, essentiality, concrete and non-frilly elegance, functional flexibility and precise idea of innovation based on the capacity to absorb and metabolize tradition. For Aldo Rossi, every project chases and reconstructs a weft of nexuses, recollections, images and memories: it is a poetic work on the traits of a prototype's permanency, where "the innovative component isn't absent, but administered in different doses and diluted in the echo of reminiscences." (Benedetto Gravagnuolo, "L'uovo che fa scoprire l'America," *Modo*, n. 103, January 1988, p. 26).

Triennale di Milano Collection

Broadway

Chair
Gaetano Pesce
1993
Bernini

A chair-spectacle: fantastical, surprising and performative like the New York area that gives it its name (and where, at 543 Broadway, the designer's New York headquarters is located). In trying to separate the notion of the "mobile" (piece of furniture) from the paradoxically "immobile," Gaetano Pesce makes a chair that is truly a *dynamic* object. If it is resting (that is if no one sits on it), *Broadway* stays still and gives an impression of relative stability. But when someone uses it, sitting on it, it immediately oscillates, flexes and sways due to legs made of springs inserted into a black nylon support. Even the color is in flux: when the epoxy resin used for the seat and back is introduced to the mould, the processing specialist is authorized to add colored pigments in quantities and varieties as desired. The result is an unrepeatable chair: just like a show on a Broadway stage, no *Broadway* model is like another, due to a very expensive process that Gaetano Pesce uses to obtain diversity in otherwise identical objects produced in a series.

Triennale di Milano Collection

Cobán

Coffee maker
Richard Sapper
1997
Alessi

It's a coffee machine, but it looks like a space station. In launching its publicity campaign it was presented as such – including a countdown – basing everything on a game of references describing it as spaceship or science fiction object that, although coming from another planet, could capture that unmistakable coffee aroma that is well-known to most of Earth's inhabitants. Nevertheless, a name with an antique flavor was given to such a "futuristic" object: Cobán is really Richard Sapper's father's birthplace, a village in upper Verapaz on the Guatemalan plateau, where the designer's grandfather had coffee plantations. The short circuit between past and future, tradition and technology, becomes the object's distinctive signature. And yet, the synergy generated by the name/shape/function nexus tends to conjure up the idea of a distance that decreases, an object that makes itself present and available for everyday use, concentrating in a room – it's of little importance whether it's a home or an office – flows of memory and flashes of imagination. Even strictly on a design level, Richard Sapper confirms a methodology in which technology, aesthetics and functionality cohabitate in an interconnected, reciprocal relationship.

Triennale di Milano Collection

Miami

Table
Thomas Sandell
1998
B&B Italia

Elegant and essential, with a structure covered in opaque aluminum chrome or white paint, the little *Miami* table was designed by Swedish designer Thomas Sandell for B&B Italia. It's characterized by its unusual top made from thermoformed Corian with a circular or oval concavity in the center – a formal but also functional detail developed to hold objects placed there, preventing them from sliding or falling. The name of Florida's capital, in this case, can sound equally arbitrary or provocative: just as Miami is a city of luxury and opulence, the small table exhibits, on the contrary, a discrete shape lacking frills and ostentation. And yet, just the name of this city actually performs on the object a perceptual mediation, inducing the user to transfer luxurious and beautiful properties (if only technological "beauty") to a product that initially seems confined to the perimeter of utility and function.

B&B Italia Collection

Rimini

Cutlery drainer
Stefano Giovannoni
1998
Alessi

Alessi Collection

The shape and color of this cutlery drainer induces immediate mnemonic references.
The visual reminder is endearing and explicit: thousands of times we have seen an old bucket abandoned on the beach together with its shovel. Thus echoes of seaside vacations, childhood games and family meals. This is Rimini, without any doubt or hesitation. Digging in the collective memory and imagination, the object's name links it to distant seaside memories and places it in the ultimate theatre of vacationing Italy. Like the Romano town sung about by Pier Vittorio Tondelli and Federico Fellini, Giovannoni's object is colorful, playful, fantastical and excessive. It tries to make the useful amusing, and to make a game of what Fellini called the most humble, convivial activity there is: the washing of dinner dishes after the party's over.

Objects/States

Compared to objects named after cities, objects named after states evidently play off a greater referential indeterminacy, both commemorative and evocative. If a city is a place that is easily visualized (for example, by appealing to a symbolic icon or an emblematic emergency), a state is much less so (if not only for the potential variety of places it contains and characterizes). This is why design objects choose the name of a state, playing on strong symbolic connotations (in the 1960s Ferrari had called a few models *America* or *California*, evidently referring to the myth of

Design by Aldo Rossi, 1975

speed and modernity on the road), or opting for relatively small and easily recognized states for a few socio-cultural characteristics. Thus, for example, Sottsass' *Yemen* vase glows with Arabian transparency, Magistretti's *Kenia* armchair plays with the nomadic quality of certain African cultures, Carlo Scarpa's *Kentucky* chair proposes a mix of modernity and tradition with an American Mid-East flavor, while the *Louisiana* armchair, again by Magistretti, and Citterio's *Panama* present an idea of comfort suspended between post-colonial indolence and elegance. Not quite a precise location in the world, these objects pursue and reconsider an idea of place, as well as suggestively communicating the impossibility of precisely visualizing and tracing its exact borders.

Kentucky

Chair
Carlo Scarpa
1979
Bernini

A striking cultural knowledge of the past and a love for refined, traditional materials are united in Carlo Scarpa's work. These qualities are combined with a polished modernity and a working methodology that prefers rationally derived, geometrically essential shapes. *Kentucky*, designed in 1979 for Bernini, reflects the exactness of this formula: in wood and leather, traced on a line that seems to want to impart in the chair an impression of reaching upward, modernity and tradition are perfectly united in a playful, refined *mélange* that unites a linear structure with a seat and backrest made of natural leather. The name, admitted to by the same designer, was chosen while leafing through a book on chairs by Shakers, a religious sect common in New England and the Mid-Eastern states of Pennsylvania, Ohio and Kentucky between the eighteenth and nineteenth centuries.

Bernini Collection

Louisiana

Armchair, footrest
Vico Magistretti
1993
De Padova

Vico Magistretti has often applied the principle of *déplacement* to his projects: for the *Sinbad* (1981) sofa he used an English horse blanket thrown upon a padded structure; for the *Incisa* (1992) armchair he transformed a leather saddle placed vertically on a steel base, using it as the main body of the chair, so that the curve of the saddle held the back of the seated person, and the lateral bands functioned as armrests. For *Louisiana*, designed a year later, again for De Padova, Magistretti returns to a similar principle. But here the natural leather color forms a rounded and plump shape that suggests – perhaps with a further effect of *déplacement* – the very name of the state from which the object takes its name. Here the comfort of an old home in New Orleans or St. Louis is evoked, its owner sunk into an armchair with his legs extended on the footrest, intently listening to the blues.

De Padova Collection

Panama

Chair
Antonio Citterio
1993
B&B Italia

Simple and essential, stripped of every frill and reduced to the archetype of a monochromatic and monomaterial chair, *Panama* (padded with removable cover) absorbs and reconsiders the off-white and tobacco tonalities linked – in both the tailor's and haberdasher's shops – to the name of the small Central American republic whose canal unites the Atlantic and the Pacific. Nothing exotic, yet in Antonio Citterio's design the evocation of a simple and functional shape with a chromatic allusion to a place becomes a structure that accentuates and underlines – in the morphology of the chair – a compositional line of great coherence and harmony.

B&B Italia Collection

Yemen

Vase
Ettore Sottsass
1994
Venini

A game of transparencies and shapes.
Classic silhouettes of geometric solids (a sphere and two truncated cones) that produce unexpected associations. A rethinking of the morphology of the vase in order to break up the most obvious design stereotypes.
In keeping with the design idea of having high emotional content and a strong communicative ability, Sottsass realizes in *Yemen* a vase that radiates colored luminescence and seems ready to furnish an interior in *A Thousand and One Nights*. Using the name of the Queen of Saba's country adds an intentionally exotic touch to the project, which and also coincides with the colors of this Middle Eastern state's flag (red/black/green).

Venini Collection

Kenia

Folding armchair
Vico Magistretti
1995
Campeggi

A simple loose covering fitted on an aluminum structure outlines the shape of this armchair that, folded, is easily carried by means of a clearly visible handle. Applying to a seat-type the same research on folding furniture already developed with the folding bookcase *Nuvola Rossa* (1977), the *Broomstick* series, and the *Tenorio* (1969) clothing stand, Magistretti now perfects a form that inspires a mobile and nomadic use, reinforced by the evident reference to African cultural practices suggested by its geographic name.

Campeggi Collection

Objects/Islands

Bruno Munari remembered that a few stones – seen from afar – look like islands. In the same way, there are also a few pieces of furniture that, observed from a special position, seem to be islands. Or, at least, the particular islands whose names they share. These are furniture-islands: artifacts that float in the home's space, sometimes combined in an archipelago, indicating a possible landing place, a shelter, a refuge or a lighthouse. It isn't unusual that the catalogue of an important historical company like Danese prefers "island" names (*Barbados, Ponza, Sicilia, Formosa,*

Design by Aldo Rossi, 1984

Canarie, Montecristo, Lampedusa): every object is basically alone and is connected with the others by a space that insists on being continuously traversed.
Whether lamps (*Falkland* and *Bali* by Munari), tables (*Lipari* by Sottsass) or vases (*Pago Pago* by Mari), objects-islands, beginning with their name, claim a sort of proud self-sufficiency or congenital autonomy. And still, at the same time, they offer themselves as possible postcard or souvenir surrogates within the home: as with certain glass bottles that contain miniature cities, these objects help preserve the memory (or dream) of the outside world, and the many islands – real or metaphorical – that make it up.

Bali

Lamp
Bruno Munari
1958
Danese

Danese Collection

A plastic cover is held rigid by two brass supports. A cube as a base supports another cube, partially rotated, that contains a light source. For Bruno Munari a lamp is conceived in this manner: from the power of an idea, from a simplicity that is difficult to force. Like a light source that floats in its surrounding environment, *Bali* conjures up the memory of Chinese puppet shadows and veiled transparencies, of Oriental dancers and lanterns in the night. The nothingness of a shape becomes the whole of the imagination, or vice versa. And Munari's method stands out like an island in Italian design's *mare magnum*.

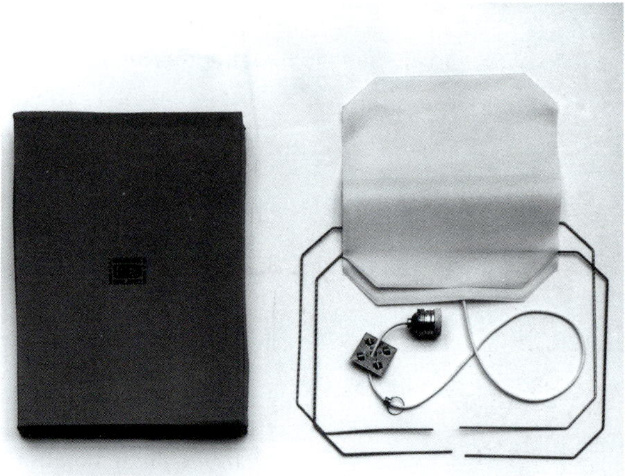

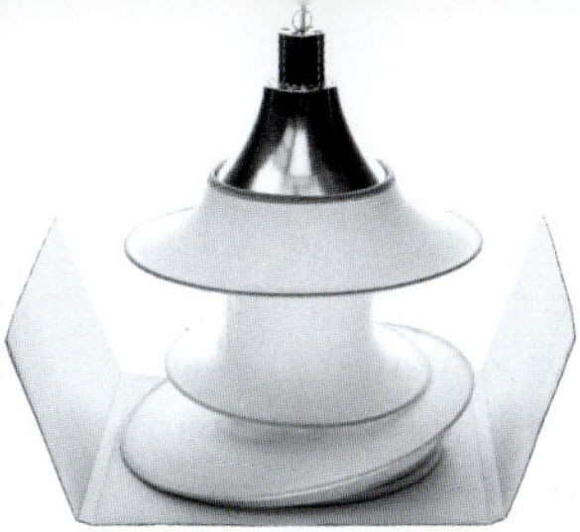

Falkland

Lamp
Bruno Munari
1964
Danese

Inspired by a woman's stocking, the *Falkland* lamp's shape is born from the tension of a tube of stretched nylon and of the weight of several metallic rings: it is a spontaneous shape, generated only by the tension of the internal forces that make it up. The white stretched nylon tube is one and a half meters long and fifteen centimeters wide. Seven metal rings, each with different diameters but always with a diameter wider than the tube, are inserted into seams set at a predetermined distance from one other. Each metal ring expands the segment of nylon tube that connects with the next ring, forming a natural curve. The entire tube with rings is hung from an aluminum reflector, which picks up the shapes of the curves with the light of a single bulb. The light filters from the tube, using the fabric's texture to create a characteristic effect of soft and indirect lighting. The name, in this case, is perhaps carefully allusive: the rings are like islands in an archipelago of light.

Triennale di Milano Collection

Lesbo

Lamp
Angelo Mangiarotti
1967
Artemide

The fruit of Angelo Mangiarotti's sculptural and organic vision, *Lesbo* mirrors the formal and functional characteristics that the designer applies to almost all of his works: the will to "concentrate all fragmentary elements which constitute an object in an organic union of sculptural continuity." (Giuliana Gramigna, *Ottagono*, n.3, October 1966, p. 77). *Lesbo,* in fact, unites in a single piece (an opaline covering) at least three different functions (that of a footrest, a column and a lighting globe), succeeding in blending in a single production practice (glass blown into a mould) the needs of serial production and the qualities of a unique and unrepeatable artisan piece. Thus each resulting *Lesbo* lamp is equal to the others, and still – at the same time – also slightly different. Also, the opaline treatment, with the coloring and softness of the material varying depending on the location, allows for the passage of more or less light at designated points. The curvilinear and sinuous design merges with the light source to make an original, sensuous design that references the Greek island with soft Sapphic shadings.

Triennale di Milano Collection

Pago Pago

Vase
Enzo Mari
1968
Danese
1997
Alessi

Playing with doubles: in this project by Enzo Mari, two vases interpenetrate each other in a single object, exactly like the two words that make up the name of the Tutuila island capital in American Samoa. Mari finds in the idea of the double vase the optimum solution for containing the cost of the mould, duplicating the holding capacity and usefulness of an artifact produced from a single matrix – all consistent with the objective of creating low-cost objects for serial production. "The aim of the research becomes to produce a complex shape, an object created around two cavities penetrating each other, so completely coherent with the injection technique that it can be easily extracted from the mould. The technical reasons connected with the extraction from the mould suggest the geometric shape of a truncated cone upon which is developed the sculptural principle of penetrating volumes." (François Burkhardt with Juli Capella and Francesca Picchi, *Perché un libro su Enzo Mari*, Federico Motta Editore, Milan 1997, p. 120). But Mari's object possesses another quality – that of an exotic island – which alters its own image according to the observer's position and viewpoint, but always maintaining an optically symmetric and mirror-like enjoyment: and like its name, the vase can be turned upside down without substantially changing its function.

Triennale di Milano Collection

Capri

Lamp, camping stove
Antonio Macchi Cassia
1981
Total gas

It's difficult to imagine "camping" in Capri – to use an object like this one on the Faraglioni island of tourism, luxury and *eat and run*. And yet, the name suggested by this *survival kit* designed by Antonio Macchi Cassia (a multi-use cooker with ledge adaptable to different sized containers and a classic camp lamp) has a provocatively ironic strength, ennobling the *do-it-yourself* vacation and at the same time mocking the rites of hedonistic tourism of the 1980s. The name really makes for a much more ambiguous and problematic object, inasmuch as it doesn't refer to its obvious function, and induces a reflection – while using it – of the many stereotypes around which the modern idea of tourism has grown and developed.

Triennale di Milano Collection

Lipari

Table
Sottsass Associates
1992
Zanotta

The ceramic, anthracite-colored base recalls the shape of a vase from *Magna Grecia*: almost an archeological find recycled for contemporary use.
The polychrome-plated top surface, in laminate, shows a decoration of concentric circles with colors that suggest the Mediterranean's brilliance and luminosity. Part of a series by Sottsass that includes various pieces of furniture all with names from the Aeolian islands (the *Vulcano* stacking table, the *Panarea* console, the *Alicudi*, *Filicudi* and *Stromboli* tables), *Lipari* offers itself as a furniture "island" within a larger archipelago of furnishings that don't demand a static location in space, but can be moved here and there according to situation, whim or necessity. The domestic space becomes an *open sea*, with furnishings as temporary landing points in the navigation of living.

Zanotta Collection

Objects/Morphologies

Like traces of nature. Like imprints, clues or traces of memory, morphological elements from the natural world forcefully enter the universe of design, suggesting lines and shapes, revealing typological models for serial production.
Vico Magistretti's *Atollo* lamp, Piero Gilardi's *I Sassi* chairs, Marco Zanuso's *Duna* silverware and Michele De Lucchi's *Lido* sofa all constitute – from this point of view – a very interesting case and an absolutely specific jump in scale: design and architecture don't connect, but – much more radically – they reunite in a *naturalism* that comes

Design by Aldo Rossi, 1984

before the design, generating a short circuit between natural and artificial, and accommodating references and allusions to that great world beyond the home within the home.
To use the *Faraglioni* as inspiration for the design of a salt and pepper mill (as in the case of Enzo Mari) or to imagine a peninsula as a table (as did Peter Shine) or even to rework the concept of island and peninsula in the structure of the kitchen environment, suggests an attempt – sometimes with affection and respect, sometimes with mimicking envy or irony – to never forget the formal lesson that nature's morphology suggests to us, to instead absorb and protect it, making it a precious and indispensable source of inspiration.

I Sassi

Seats
Piero Gilardi
1968
Gufram

Seen from very far away some stones are like small worlds with their own continents, islands and seas. This is how astronauts see planet Earth wrapped in clouds. (Bruno Munari, *Da lontano era un'isola*, Edizioni E. Elle, Einaudi, 1994, p. 7).

Or rather, seen from up close, Piero Gilardi's *I Sassi* (the stones) is a series of soft seats, completely made of polyurethane, part of the so called "mimic project" produced by Gufram during the 1980s (*Pratone*, *Bocca*, *Capitello*, *Cactus*, etc.) that hovered between irony and radicalism.
I Sassi introduces the game of binomial opposition (true-false, hard-soft, heavy-light, exterior-interior) to the domestic space and to the world of serial production, provoking a perceptive reversal in the user and dissolving the need make clear borders separating *natural* and *artificial*.

Triennale di Milano Collection

Faraglioni

Salt and pepper mill
Enzo Mari
1972
Danese
1997
Alessi

They have the same shape as the stone posts designed by Mari as urban barriers furnishing the streets of Milan, but obviously they have a different function. And yet, in this game of breaking down relations between shape and function, we find in this formal and analogical union of functionally different elements (destined for both the macro urban scale as well as the micro domestic scale) one of the more specific peculiarities of Mari's work.
Similar to rocks emerging from the sea, with plunging sides and rounded bases, the *Faraglioni* salt and pepper mills transform nature's shapes into functional design elements: the skilful use, for example, of grooves on the object's body – elements that characterize the natural morphology of the design's form – here become specific functional traits that make up a salt and pepper mill.

Alessi Collection

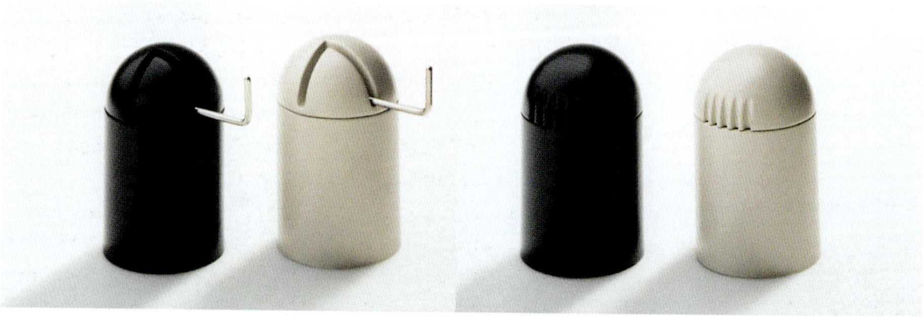

Atollo

Lamp
Vico Magistretti
1977
O-Luce

Triennale di Milano Collection

A cylinder, a cone and a hemisphere. Three solid Euclidian shapes, returned to their geometric and formal purity, comprise the structural and compositional elements of this lamp, awarded the Compasso d'Oro (Golden Compass) in 1979.
Seen frontally, the hemisphere floats in a surprising equilibrium on the cone's vertex, bringing to mind certain American minimalist sculptures.
But *Atollo* (atoll) isn't only a question of shape. It's also an experiment with light: Magistretti positions the light source inside the spherical segment, allowing it to appear completely shadowed from the outside, while the interior and metal point are violently lit, tangentially bathing the shaft-cylinder with light. The effect is one of light acting on an environment with different reflective intensities, propagating itself directly and indirectly. Like an island of light immersed in a sea of shadow, *Atollo* synthesizes the lamp-prototype with a perfection that makes its abstract shapes come alive, giving a new meaning to the natural morphology evoked by the name.

Lido

Sofa
Michele De Lucchi
1982
Memphis

One lands on a beach (lido), one relaxes and rests on a beach (lido). This is how, by transferring on the sofa-type the typical connotations of the natural morphology conjured up by the object's name, Michele De Lucchi – in keeping with the Memphis mission to make design capable of being above all communicative while stirring up emotional reactions – works in an almost philological manner on the idea of *flatness*, *horizontality* and *stratification* (a lido is typically defined as a strip of flat land near the sea or as a stratified deposit/accumulation of sandy or pebbly material). In wood, laminated plastic and metal, *Lido* also uses colors to emulate the gaudy and flashy colors that appear when a lido is transformed into a bathing location, combining in the object the idea of relaxation with that of worldliness.

Memphis Collection

Peninsula

Side table
Peter Shire
1982
Memphis

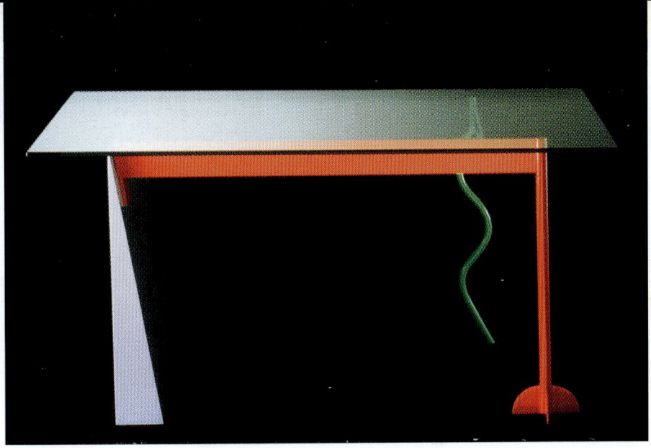

Asymmetric. Sculptural. Postmodern. The adjectives used to define the work of Californian designer Peter Shire – called in by Sottsass to be part of the Memphis group – also adapt well to the style of *Peninsula*, the metal and crystal side table that – as a diptych with *Brazil* of 1981 – revisits one of the most experimented with and codified pieces of furniture in the name of decidedly *funky* taste.
Warm colors, mixed materials and absolutely non-functional lines. With elements that stretch out in an overhanging penetration of space, *Peninsula* demolishes the traditional symmetry of table legs (two plus a solid volume instead of the usual four), adopting curved lines instead of straight ones and using color as a further chromatic differentiation.
Variety replaces uniformity, in an artifact that – although elements recall a natural morphology – is antiphrastically formed as an emblem of the most absolute artificiality.

Memphis Collection

Duna

Silverware service
Marco Zanuso
1995
Alessi

These pieces of silverware need to be admired in profile with their bending, curving and gradually tapering structure, in a game of curves and contours that recalls, by analogy, the silhouette of a dune-covered desert landscape.
Returning to a design presented at the *Silver Design Competition in Italy*, publicly advertised in 1959 by the American company Reed & Barton, Marco Zanuso designed elegantly simple shapes, including a fork that flows and lengthens along the handle, and with a slightly deeper central space between the two inner tines.
Alberto Alessi remembers that Zanuso singled out the central characteristic element of these pieces as their "small waist," or rather, the "narrowing of the end part of the handle, really like the dresses of the '50s worn by girls from good families." (Alberto Alessi, *La fabbrica dei sogni*, Electa/Alessi, Milan 2002, p. 82). It's an evocative reference, especially when referring to a single piece. But seen all together, from the side or from above, this silverware series in fact evokes the morphology of a landscape inserted into the conviviality of a domestic interior.

Alessi Collection

Setting Up

Giancarlo Basili

Giancarlo Basili (1952) is one of the most important set designers in Italian cinema. He has worked with, among others, Gabriele Salvatores (*Nirvana*), Gianni Amelio (*Cosi ridevano*) and Nanni Moretti (*Palombella Rossa, La Stanza del figlio*).

The World in a Room is the third exhibition in the *Words and Things* series.

The space that held these three exhibitions has always been the same: the Impluvium on the first floor of the Triennale di Milano's Palazzo dell'Arte. For me, working for three consecutive periods in the same space has been an absolutely new experience: like a series, or a sequel, I tried to translate the concepts expressed in each show by working on the variations of sameness, or on finding the analogy within the difference.

In this case, it is about the idea of a world made

up of different objects being returned to the inside of a room. I felt the need to widen and alter the space in order to create new frontiers, new horizons and new perspectives. I had painted the walls in white and black, with lines of perspective produced by different focal points, following the technique of optical art; I covered the floor with mirrored material in order to reflect depth and to view the objects upside down. Even the two walls of the niches were covered with mirrors, so as to created a multiplication of

rooms, or to create the illusion of a *via novissima*, of a rarefied, abstract, suspended space, in order to host the muddle from the outside world.

Objects with Names of Places

Essential Index

Abitacolo · shelving system · 1971 · Bruno Munari · Robots
Acapulco · salt and pepper shakers · 1959 · Gino Colombini · Kartell
Alaska · vase · 1982 · Ettore Sottsass · Memphis
Albenga · small armchair · 1955 · Gustavo F. Pulitzer · Arflex
Alicudi · table · 1992 · Sottsass Associati · Zanotta
Ameland · letter opener · 1962 · Enzo Mari · Danese
America · car · 1951 · Ferrari
Arcadia · lamp · 1994 · Gismondi e Fassina · Artemide
Arcipelago (archipelago) · kitchen unit · 1972 · Giancarlo Iliprandi · RB Rossana
Arizona · fabric · 1983 · Nathalie Du Pasquier · Memphis
Asolo · table · 1980 · Angelo Mangiarotti · Skipper
Assuan · armchair · 2000 · Vico Magistretti · De Padova
Atlantic · chest of drawers · 1981 · Michele De Lucchi · Memphis
Atollo (atoll) · lamp · 1977 · Vico Magistretti · O-Luce
Babele (Babel) · étagère · 1987 · Massimo Morozzi · Fiam Italia
Baia (Baja) · seat · 1975 · Citterio and Nava · B&B Italia
Bali · lamp · 1958 · Bruno Munari · Danese
Bali · magazine rack · 1993 · K. Krizek · Zanotta
Barbados · ashtray · 1964 · Angelo Mangiarotti · Danese
Basilica (la) · table · 1976 · Mario Bellini · Cassina
Bel Air · seat · 1982 · Peter Shire · Memphis
Bellagio · armchairs and sofas · 1961 · Sergio Asti · Adrasteia
Berlino (Berlin) · centerpiece · 1998 · Enzo Mari · Sawaya e Moroni
Beverly · cabinet · 1981 · Ettore Sottsass · Memphis
Bisanzio (Byzantium) · sofa · 1988 · Alessandro Mendini · Zanotta
Bombay · small table · 1986 · Nathalie Du Pasquier · Memphis
Bordeaux · lamp · 1986 · Nathalie Du Pasquier · Memphis
Borgogna (Burgundy) · armchair · 1964 · Cini Boeri · Arflex
Borneo · ashtray · 1967 · Enzo Mari · Danese
Boston · cocktail set · 1979 · Ettore Sottsass · Alessi
Brasilia · coffee maker · 1962 · Ponti, Fornaroli, Rosselli · La Pavoni
Brazil · end table · 1981 · Peter Shire · Memphis
Brera · lamp · 1992 · Achille Castiglioni · Flos
Broadway · chair · 1993 · Gaetano Pesce · Bernini
Brooklyn · bookshelf · 1977 · Acerbis and Stoppino · Acerbis International
Buenos Aires · lamp · 1986 · Cibic and Ongaro · Memphis
Burundi · fabric · 1981 · Nathalie Du Pasquier · Memphis
Cabina dell'Elba (cabin of Elba) · closet · 1980 · Aldo Rossi · Molteni
Cairo · small table · 1986 · Michele De Lucchi · Memphis
California · car · 1966 · Ferrari
California · sofa · 1968 · Edelhard Harris · ICF
California · motorcycle · 1975 · Guzzi
California · fabric · 1983 · Nathalie Du Pasquier · Memphis
Canada · sofa · 1962 · Ettore Sottsass · Poltronova
Canarie (Canary Islands) · desk set · 1958 · Bruno Munari · Danese
Cannaregio · modular sofa · 1987 · Gaetano Pesce · Cassina
Capitello (capital) · armchair · 1970, Studio 65 · Gufram
Capitello (capital) · table · 1977 · Enzo Mari · Driade spa
Capri · lamp · 1961 · Bruno Munari · Danese
Capri · lamp, camping stove · 1981 · Antonio Macchi Cassia · Total Gas
Carimate · chair · 1960 · Vico Magistretti · Cassina
Casablanca · sideboard · 1981 · Ettore Sottsass · Memphis
Castello (il) (the castle) · seats · 1976 · Mario Bellini · Cassina
Cattedrale (la) (the cathedral) · chair prototype · 1968 · Afra and Tobia Scarpa · B&B Italia
Ceylon · umbrella stand · 1962 · Enzo Mari · Danese
Chelsea · bed · 1981 · George Sowden · Memphis
Chester · armchair · 1912 · Renzo Frau · Chair Frau
Cipro (Cyprus) · ashtray · 1974 · Bruno Munari · Danese
Cnosso · lamp · 1969 ·

Angelo Mangiarotti · Artemide
Cobán · coffee maker · 1997 · Richard Sapper · Alessi
Colonnato (il) (the colonnade) · table · 1977 · Mario Bellini · Cassina
Colorado · teapot · 1983 · Marco Zanini · Memphis
Corinto (Corinth) · lamp · 1993 · Ernesto Gismondi · Artemide
Corinto (Corinth) · pots · 1948/86 · E. Barovier · Barovier e Toso
Corsica · desk set · 1962 · Bruno Munari · Danese
Corte (la) (the court) · seats · 1976 · Mario Bellini · Cassina
Cupola (la) · coffee pot · 1989 · Aldo Rossi · Alessi
D'Antibes · china cabinet · 1981 · George J. Sowden · Memphis
Delfi (Delphi) · table · 1970 · C. Scarpa, M. Breur · Simon
Delos · ashtray · 1981 · Enzo Mari · Danese
Dublin · sofa · 1981 · Marco Zanini · Memphis
Duna (dune) · silverware service · 1995 · Marco Zanuso · Alessi
Egina (Aegina) · ceiling light · 1979 · Angelo Mangiarotti · Artemide
Elba · umbrella stand · 1961 · Enzo Mari · Danese
Eldorado · motorcycle · 1972 · Guzzi
Europa (Europe) · car · 1953 · Ferrari
Europa (Europe) · sofa · 1988 · Gualtierotti, Mazzoni delle Stelle · Zanotta
Falkland · lamp · 1964 · Bruno Munari · Danese
Faraglioni · salt and pepper mill · 1972 · Enzo Mari · Danese, Alessi
Festo · table · 1972 · Gae Aulenti · Zanotta
Filicudi · table · 1992 · Sottsass Associati · Zanotta
Firenze (Florence) · small armchair · 1967 · Pierluigi Spadolini · Arflex
Firenze (Florence) · lamp · 1994 · Ettore Sottsass · Venini
Firenze (Florence) · clock · 1996 · Achille Castiglioni · Alessi
Florence · armchair · 1995 · Antonio Citterio · B&B Italia
Florence · furniture · 1999 · A. Haberli · Zanotta
Florida · motorcycle · 1989 · Guzzi
Formosa · calendar · 1963 · Enzo Mari · Danese
Gabon · carpet · 1982 · Nathalie Du Pasquier · Memphis
Gibilterra (Gibraltar) · vase · 2002 · Andrea Branzi · Alessi
Ginevra (Geneva) · chair · 1979 · Achille Castiglioni · BBB Bonacina
Gloucester · small armchair · 1986 · George J. Sowden · Memphis
Granada · lamp · 1967 · Luciano Buttura · Guzzini
Grand'Italia · bench and dormeuse · 1979 · De Pas, D'Urbino, Lomazzi · Zanotta
Graz · table · 2001 · Sawaya · Sawaya e Moroni
Guadalupa · ashtray · 1972 · Enzo Mari · Danese
Havana · small armchair · 1994 · Pietro Arosio · Alivar Gallery
Hollywood · small table · 1983 · Peter Shire · Memphis
Isola (island) · furniture unit · 1978 · A. Rossi · Bernini
Isola (island) · kitchen/dining room unit · 1986 · Damiano Alberti · Alberti
Java · table/container · 1968 · Enzo Mari · Danese
Kenia (Kenya) · folding chair · 1995 · Vico Magistretti · Campeggi
Kentucky · chair · 1979 · Carlo Scarpa · Bernini
Kerguelen · coatrack · 1967 · Enzo Mari · Danese
Kyoto · low table · 1974 · Gianfranco Frattini · Ghianda
Kyoto · small table · 1983 · Shiro Kuramata · Memphis
Lambro · three-wheeled vehicle · 1948 · Pier Luigi Torre · Innocenti
Lampedusa · pencil case · 1967 · Enzo Mari · Danese
Le Mans · motorcycle · 1984 · Guzzi
Lesbo · lamp · 1967 · Angelo Mangiarotti · Artemide spa
Lido (beach) · sofa · 1982 · Michele De Lucchi · Memphis
Lierna · chair · 1960 · Achille and Pier Giacomo Castiglioni · Cassina, Gavina
Lima · seat · 1995 · Jasper Morrison · Cappellini
Linz · lamp · 1977 · Vico Magistretti · O-Luce

Luca Meda · Molteni

Ponte (bridge) · small table · 1981 · A. Cortesi · Fiam Italia

Ponza · ashtray · 1958 · Bruno Munari · Danese

Rialto · furniture · 1977 · Carlo Scarpa · Sinan

Rimini · cutlery drainer · 1998 · Stefano Giovannoni · Alessi

Riviera · chair · 1981 · Michele De Lucchi · Memphis

Roma (Rome) · small armchair · 1986 · Marco Zanini · Memphis

Rotonda (la) (the rotunda) · table · 1976 · Mario Bellini · Cassina

Sahara · motorcycle · 1977 · Guzzi

Salina (salt flat) · desk set · 1985 · Enzo Mari · Danese

Samarcanda · seats · 2001 · Vico Magistretti · De Padova

Samos · baskets · 1973 · Enzo Mari · Danese

Sanremo · lamp · 1968 · Archizoom · Poltronova

Sanremo · chair · 1975 · De Pas, D'Urbino, Lomazzi · Zanotta

Santa Ana · lamp · 1983 · Matteo Thun · Memphis

Santa Fe · lamp · 1983 · Matteo · Thun · Memphis

Santa Monica · lamp · 1983 · Matteo Thun · Memphis

Sassi (I) (the stones) · seats · 1968 · Piero Gilardi · Gufram

Segesta · mirror · 1989 · Matteo Thun · Kartell

Sicilia (Sicily) · basket · 1959 · Bruno Munari · Danese

Soho · furniture · 2001 · Emaf Progetti · Zanotta

St. Moritz · small bed · 1951 Marco Zanuso · Arflex

Stromboli · small table · 1993 · Sottsass Associati · Zanotta

Suomi · vase · 1998 · Mendini and Rasulo · Zanotta

Sverige · vase · 1998 · Mendini and Rasulo · Zanotta

Tahiti · lamp · 1981 · Ettore Sottsass · Memphis

Tamigi (Thames) · sofa · 1988 · Zanotta Archivio · Zanotta

Tanganika · chair · 1979 · Vico Magistretti · Alias

Tangeri (Tangiers) · sofa · 1994 · Francesco Binfarè · Edra

Tatlin · seat · 1989 · Cananzi and Semprini · Edra

Teatro (theater) · small armchair · 1982 · Meda and Rossi · Molteni

Terracina · vase · 2002 · Andrea Branzi · Alessi

Ticino · sofa · 1986 · Zanotta Archivio · Zanotta

Timor · calendar · 1967 · Enzo Mari · Danese

Tokio (Tokyo) · lamp · 1980 · Asahara Sigheaki · Stilnovo

Toledo · bed · 1975 · Carlo Scarpa · Sinan

Top Kapi · sofa · 1998 · Francesco Binfarè · Cassina

Topkapi · vase set · 1997 · Guggisberg Baldwin · Venini

Torcello · furniture · 1965 · Afra and Tobia Scarpa · Stildomus

Tramonto a New York (sunset in New York) · sofa · 1980 · Gaetano Pesce · Cassina

Trieste · chair · 1961 · Guglielmo Ulrich · Saffa

Trieste · chair · 1966 · Pierangelo D'Aniello, Aldo Jacober, Bazzani

Trumeau · sideboard · 1951-1970 · Piero Fornasetti and Gio Ponti/1980 · Immaginazione srl

USA · vase · 1998 · Mendini and Rasulo · Zanotta

Venezia (Venice) · vases · 1999 · Barovier e Toso

Ventotene · desk set · 1962 · Enzo Mari · Danese

Vesuvio · vases · 1999 · R. Plesl · Barovier e Toso

Vienna · table · 1979 · Peter Noever · Zanotta

Vienna · table/storage unit · 2000 · Antonio Citterio · Flexform

Volterra · vase · 2002 · Andrea Branzi · Alessi

Vulcano · vases · 2000 · S. Malik · Barovier e Toso

Yemen · vase · 1994 · Ettore Sottsass · Venini

Zaire · fabric · 1982 · Nathalie Du Pasquier · Memphis

Zambia · fabric · 1982 · Nathalie Du Pasquier · Memphis

Zurigo (Zurich) · sofa · 1998 · Haberli, Marchand · Zanotta

We thank the following
Alessi · Artemide · B&B Italia · Bernini · Campeggi · Cassina · Danese · De Padova · Edra · Fiam Italia · Fontana Arte · Memphis · Poltrona Frau · Poltronova · Venini · Zanotta

For their kind collaboration
Francesca Appiani · Alberto Bianchi Albrici · Maristella Bonalumi · Claudio Campeggi · Francesca Cavazzuti · Isabella Colombo · Roberta Meloni · Luciana Motta · Maria Laura Rossiello · Giovanna Solinas · Raffaella Zuccarini

and also
Matteo Dolcetti · Enrico Mastrapasqua · Cinzia Prestigiacomo

Photographic Credits

Archivio fotografico della Triennale, Milano,
Ph. by Amendolagine and Barracchia: pp. 36, 37, 55, 61, 62, 64, 67, 97

Archivio Alessi: pp. 34, 50 (Ph. Riccardo Bianchi), 51, 56, 63 (Ph. Aldo Ballo), 70 (Ph. Riccardo Bianchi), 73 (Ph. Carlo Lavatori), 82, 89 (Ph. Santi Caleca), 92, 96 (Ph. Santi Caleca), 101
Archivio Artemide: p. 87
Archivio B&B Italia: pp. 71, 78
Archivio Bernini: pp. 69, 76
Archivio Campeggi: p. 81
Archivio Cassina: pp. 43 (Ph. Aldo Ballo), 59, 65 (disegno di G. Pesce)
Archivio Danese: pp. 84, 85
Archivio De Padova: p. 77
Archivio Edra: p. 53
Archivio Fiam Italia: pp. 47, 49
Archivio Fontana Arte: p. 38
Archivio Gufram: p. 95
Archivio A. Macchi Cassia: p. 90
Archivio Memphis: pp. 98, 99
Archivio Molteni&C: p. 68
Archivio Poltrona Frau: p. 58
Archivio Venini: p. 79
Archivio Zanotta: p. 91 (Ph. Ramazzotti)

p. 55 from: Sergio Polano, *Santiago Calatrava. Opera completa*, Electa, 1996
pp. 44, 45 from: Aldo Rossi, *Le cabine dell'Elba e altri mobili*, Bruno Longoni, Milano 1993
p. 49 (Pieter Bruegel the Elder, painting, 1563) from: "Rassegna," n. 16, December 1983

Printed in November 2002
by La Grafica-Cantù, e-mail: lagrafica@cracantu.it
for Edizioni Charta